Assessment

Douglas H. Clements

·

Julie Sarama

Education

Bothell, WA • Chicago, IL • Columbus, OH • New York, NY

This curriculum was supported in part by the National Science Foundation under Grant No. ESI-9730804, "Building Blocks-Foundations for Mathematical Thinking, Pre-Kindergarten to Grade 2: Research-based Materials Development" to Douglas H. Clements and Julie Sarama. The curriculum was also based partly upon work supported in part by the Institute of Educational Sciences (U.S. Dept. of Education, under the Interagency Educational Research Initiative, or IERI, a collaboration of the IES, NSF, and NICHHD) under Grant No. R305K05157, "Scaling Up TRIAD: Teaching Early Mathematics for Understanding with Trajectories and Technologies" and by the IERI through a National Science Foundation NSF Grant No. REC-0228440, "Scaling Up the Implementation of a Pre-Kindergarten Mathematics Curricula: Teaching for Understanding with Trajectories and Technologies." Any opinions, findings, and conclusions or recommendations expressed in this material are those of the authors and do not necessarily reflect the views of the funding agencies.

www.mheonline.com

Send all inquiries to:
McGraw-Hill Education
8787 Orion Place
Columbus, OH 43240

ISBN: 978-0-02-127266-2
MHID: 0-02-127266-2

Printed in the United States of America.

2 3 4 5 6 7 8 9 DOH 18 17 16 15 14 13

Table of Contents

Building Blocks Assessment

Building Blocks is rich in opportunities and resources to conduct comprehensive assessments that inform instruction. The *Building Blocks Assessment* is designed to evaluate all math proficiencies. Goals of assessment are to improve instruction by informing teachers about the effectiveness of their lessons, promote growth of children by identifying where they need additional instruction and support, and recognize accomplishments.

Building Blocks assessments have the following characteristics to ensure reliable feedback:

- **Rich Variety** Assessments cover all five proficiencies: understanding, reasoning, computing, applying, and engaging, so teachers can assess a rich variety of mathematics topics and problem situations.

- **Multiple Sources** A number of assessment opportunities are included in the program so teachers gather evidence from many different sources.

- **Alignment** Assessments are carefully aligned with the curriculum and instruction and test what has been covered. The curriculum is correlated with national and state standards so the assessments cover the required content.

Summative Assessments and Differentiating Instruction
At the end of each week, teachers review children's progress in mathematics by looking at the Weekly Record Sheets (Monday, Wednesday, Friday) and the Small Group Record Sheets (Tuesday, Thursday) from the past week.

Using *Online Assessment*, an electronic assessment tool for recording and tracking children's progress, teachers summarize and analyze assessment data for each child based on the weekly observations and Record Sheets. Such information helps determine where each child is on the math trajectory. Once a child exhibits specific levels of the trajectory, begin to encourage and work with that child toward the next level. Refer to the appendix in the *Teacher's Edition* for individualized instruction opportunities, including Special Education.

Building Blocks software management system tracks children's progress on the learning trajectories through the software activities. The *Building Blocks* software provides reports to compare data and share with children and their families.

Types of Assessment

Building Blocks provides a variety of ready-made assessments to help teachers determine what children know to inform instruction.

Simple record sheets enable teachers to record and track their observations of children, which can later be recorded using *Online Assessment* to help provide a more complete view of children's proficiency.

Weekly Record Sheets

Use the Weekly Record Sheets to record children's participation and progress on math activities.

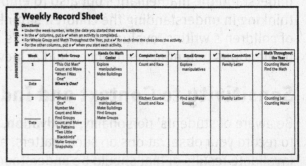

Small Group Record Sheets

Small Group Record Sheets help track each child's participation in Small Group activities. Because of their special nature, these activities are an effective and convenient means of monitoring children.

Learning Trajectory Records

These assessments are specifically tailored to provide information on where a child may be in each learning trajectory. Teachers who understand the levels of the trajectories are much more effective than others in providing appropriate feedback and activities to develop their children's math proficiency.

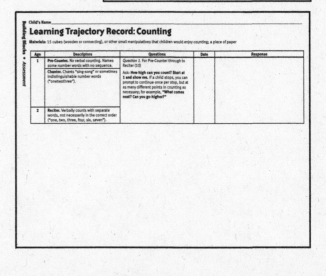

Analyzing Children's Work

A key factor in assessing children's work involves mathematical problem solving on the part of the teacher. It is often easy to identify that an answer is wrong or correct. On the other hand, it is often difficult to understand why an answer is incorrect, how the child arrived at the wrong answer, and what can be done about it. This evaluation requires the teacher to not only understand the mathematics but also to employ both critical and creative thinking in understanding the children's thinking. As in thorough evaluation of children's writing, evaluation of children's mathematical thinking has tremendous potential to advance child achievement.

Sensitivity to Context and Individuals

Be aware of students' personality, motivation, and attention. Make notes to record your observations on such matters, especially those that may invalidate results or that should be understood to protect the rights of the student. For example, if a child has a disability, especially a severe disability such as blindness or lack of movement, which interferes with the perception of assessment items or ability to respond to them, results may not be valid unless effectivce adaptations of the assessment have been made to accommodate the child. With all students, but especially with young students and students with special needs, sensitivity to contexts and individuals is needed, and principled recommendations that no single assessment be interpreted in isolation from other information should be heeded.

Weekly Record Sheet

Directions
- Under the week number, write the date you started that week's activities.
- In the ✔ columns, put a ✔ when an activity is completed.
- For Whole Group and Math Throughout the Year, put a ✔ for each time the class does the activity.
- For the other columns, put a ✔ when you start each activity.

Week	✔	Whole Group	✔	Hands On Math Center	✔	Computer Center	✔	Small Group	✔	Home Connection	✔	Math Throughout the Year
1 ___ Date		"This Old Man" Count and Move "When I Was One" *Where's One?*		Explore manipulatives Make Buildings		Count and Race		Explore manipulatives		Family Letter		Counting Wand Find the Math
2 ___ Date		"When I Was One" Number Me Count and Move Find Groups Count and Move in Patterns "Two Little Blackbirds" Make Groups Snapshots		Explore manipulatives Make Buildings Find Groups Make Groups		Kitchen Counter Count and Race		Find and Make Groups		Family Letter		Counting Jar Counting Wand

Building Blocks • *Assessment*

1

Weekly Record Sheet

2

Directions:
- Under the week number, write the date you started that week's activities.
- In the ✔ columns, put a ✔ when an activity is completed.
- For Whole Group and Math Throughout the Year, put a ✔ for each time the class does the activity.
- For the other columns, put a ✔ when you start each activity.

Week	✔	Whole Group	✔	Hands On Math Center	✔	Computer Center	✔	Small Group	✔	Home Connection	✔	Math Throughout the Year
3 Date		"Baker's Truck" Compare Number Pizzas Count and Move in Patterns Demonstrate Counting Number Me (5) Counting Book		Find the Number Fill and Spill Draw Numbers		Pizza Pizzazz 1		Make Number Pizzas Demonstrate Counting		Family Letter		Counting Wand (Count All) Snack Time Simon Says Numbers
4 Date		Count and Move in Patterns Match and Name Shapes Match Blocks Circle Time! *Building Shapes* Circles and Cans "Wake Up, Jack- in-the-Box" "Circle" Circle or Not?		Explore Shape Sets Match Shape Sets Circles and Cans		Mystery Pictures 1		Match and Name Shapes		Family Letter		Counting Wand Dough Shapes Foam Puzzles

Weekly Record Sheet

Directions:
- Under the week number, write the date you started that week's activities.
- In the ✔ columns, put a ✔ when an activity is completed.
- For Whole Group and Math Throughout the Year, put a ✔ for each time the class does the activity.
- For the other columns, put a ✔ when you start each activity.

Week	✔	Whole Group	✔	Hands On Math Center	✔	Computer Center	✔	Small Group	✔	Home Connection	✔	Math Throughout the Year
5 Date ___		***Building Shapes*** Shape Show "Wake Up, Jack-in-the-Box" Snapshots		Shape Hunt Straw Shapes		Mystery Pictures 2 Number Snapshots 1		Straw Shapes Is It or Not?		Family Letter		Name Faces of Blocks Shape Walk
6 Date ___		"Baker's Truck" Make Number Pizzas Places Scenes Count and Move in Patterns ***Where's One?*** Snapshots Number Me "Five Dancing Dolphins"		Make Number Pizzas Find the Number Draw Numbers Places Scenes		Pizza Pizzazz 2 Road Race		Pizza Game 1 Make Number Pizzas		Family Letter		Simon Says Numbers Snack Time

Building Blocks • *Assessment*

3

Weekly Record Sheet

Directions:
- Under the week number, write the date you started that week's activities.
- In the ✔ columns, put a ✔ when an activity is completed.
- For Whole Group and Math Throughout the Year, put a ✔ for each time the class does the activity.
- For the other columns, put a ✔ when you start each activity.

Week	✔	Whole Group	✔	Hands On Math Center	✔	Computer Center	✔	Small Group	✔	Home Connection	✔	Math Throughout the Year
7 Date ___		Numeral 1 *Goldilocks and the Three Bears* Numeral 2 Compare Snapshots Numeral 3 Number Jump Numeral 4 Numeral 5		*Goldilocks and the Three Bears* Pizza Game 1 Places Scenes		Party Time 1 Road Race: Shape Counting		Get Just Enough Compare Game		Family Letter		Numerals Every Day Setting the Table
8 Date ___		Numeral 6 Listen and Count Listen and Copy Where's My Number? Number Jump		Compare Game Get Just Enough Find the Number Places Scenes Pizza Game 1		Numeral Train Game Pizza Pizzazz 3		Number Jump Number Race		Family Letter		Counting Jar Numerals Every Day

4

Building Blocks • *Assessment*

Weekly Record Sheet

Directions:
- Under the week number, write the date you started that week's activities.
- In the ✔ columns, put a ✔ when an activity is completed.
- For Whole Group and Math Throughout the Year, put a ✔ for each time the class does the activity.
- For the other columns, put a ✔ when you start each activity.

Week	✔	Whole Group	✔	Hands On Math Center	✔	Computer Center	✔	Small Group	✔	Home Connection	✔	Math Throughout the Year
9 ___ Date		Listen and Copy Shape Flip Book "Three Straight Sides" Listen and Count Is It or Not? Number Jump Rectangles and Boxes Count and Move in Patterns Number Jump (Numerals)		Shape Flip Book Is It or Not? Match Shape Sets Rectangles and Boxes Compare Game		Memory Geometry 1 Number Snapshots 2		Is It or Not? Memory Geometry		Family Letter		Counting Jar Name Faces of Blocks Numerals Every Day Shape Walk
10 ___ Date		I Spy Shape Step "Five Little Fingers" Guess My Rule *Building Shapes*		Shape Pictures Memory Geometry Shape Flip Book Compare Game Rectangles and Boxes		Mystery Pictures 2 Memory Geometry 2		Guess My Rule Shape Step		Family Letter		Counting Jar Numerals Every Day Shape Walk

Building Blocks • *Assessment*

5

Weekly Record Sheet

Directions:
- Under the week number, write the date you started that week's activities.
- In the ✔ columns, put a ✔ when an activity is completed.
- For Whole Group and Math Throughout the Year, put a ✔ for each time the class does the activity.
- For the other columns, put a ✔ when you start each activity.

Week	✔	Whole Group	✔	Hands On Math Center	✔	Computer Center	✔	Small Group	✔	Home Connection	✔	Math Throughout the Year	✔
11 Date ___		Number Jump (Numerals) Listen and Count How Many Now? "Little Bird" Numeral 7 Numeral 8 How Many Now? (Hidden version)		Memory Number Places Scenes Compare Game Pizza Game 1		Party Time 2 Memory Number 1		How Many Now? Number Choice		Family Letter		Cleanup (Pick a Number) Count Motions (Transitioning) Counting Jar Numerals Every Day	
12 Date ___		Numeral 9 Number Jump (Numerals) *Makayla's Magnificent Machine* Mr. Mixup "Ten Little Birdies" Numeral 10		Dinosaur Shop Places Scenes Memory Number Numeral Sorting		Dinosaur Shop 1 Space Race: Number Choice		Mr. Mixup Dinosaur Shop (Sort and Label)		Family Letter		Dough Numerals Dinosaur Shop (Dramatic Play) Numerals Every Day	

Weekly Record Sheet

Directions:
• Under the week number, write the date you started that week's activities.
• In the ✔ columns, put a ✔ when an activity is completed.
• For Whole Group and Math Throughout the Year, put a ✔ for each time the class does the activity.
• For the other columns, put a ✔ when you start each activity.

Week	✔	Whole Group	✔	Hands On Math Center	✔	Computer Center	✔	Small Group	✔	Home Connection	✔	Math Throughout the Year
13 ____ Date		Count and Move Build Cube Stairs Count and Move (Forward and Back) *Victor Diego Seahawk's Big Red Wagon* Order Cards "Five Little Monkeys"		Build Cube Stairs Dinosaur Shop Places Scenes Order Cards		Build Stairs 1 Build Stairs 2		Build Cube Stairs How Many Now? (Hidden version)		Family Letter		Cleanup Dinosaur Shop (Dramatic Play) Numerals Every Day
14 ____ Date		Shape Flip Book Trapezoids Count and Move (Forward and Back) Feely Box (Match) *Building Shapes* Shape Step Feely Box (Name) Computer Show		Shape Flip Book Shape Pictures Feely Box (Name) Shape Step		Mystery Pictures 3 Memory Geometry 3		Feely Box (Match and Name)		Family Letter		Counting Jar Dinosaur Shop (Dramatic Play) Guessing Bag Shape Walk

Building Blocks • *Assessment*

7

Weekly Record Sheet

Directions:
- Under the week number, write the date you started that week's activities.
- In the ✔ columns, put a ✔ when an activity is completed.
- For Whole Group and Math Throughout the Year, put a ✔ for each time the class does the activity.
- For the other columns, put a ✔ when you start each activity.

Week	✔	Whole Group	✔	Hands On Math Center	✔	Computer Center	✔	Small Group	✔	Home Connection	✔	Math Throughout the Year
15 ___ Date		Shape Step Mr. Mixup (Shapes) Count and Move (Forward and Back) Guess My Rule How Many Now? Discuss Shape Pictures		Shape Pictures Feely Box (Name) Shape Flip Book		Mystery Pictures 4 Memory Geometry 4 Memory Geometry 5		Guess My Rule Shape Step		Family Letter		Counting Jar Shape Books Shape Walk
16 ___ Date		Dancing Patterns Pattern Strips Count and Move in Patterns Listen and Copy Stringing Beads Computer Show		Pattern Strips Stringing Beads		Pattern Zoo 1 Pattern Planes 1 Marching Patterns 1		Pattern Strips		Family Letter		Clothes Patterns Creative Pattern People Patterns

Weekly Record Sheet

Directions:
- Under the week number, write the date you started that week's activities.
- In the ✔ columns, put a ✔ when an activity is completed.
- For Whole Group and Math Throughout the Year, put a ✔ for each time the class does the activity.
- For the other columns, put a ✔ when you start each activity.

Week	✔	Whole Group	✔	Hands On Math Center	✔	Computer Center	✔	Small Group	✔	Home Connection	✔	Math Throughout the Year
17 Date ___		"Oh Dear, What Can the Pattern Be?" Pattern Strips (The Core) Count and Move in Patterns Cube Patterns Listen and Copy		Pattern Strips (The Core) Stringing Beads Build Cube Stairs Cube Patterns		Pattern Zoo 2 Pattern Planes 2 Marching Patterns 2		Pattern Strips (The Core)		Family Letter		Clothes Patterns Pattern Walk People Patterns Playground Patterns
18 Date ___		Number Jump (Numerals) Snapshots Listen and Copy Guess My Rule Mr. Mixup (Shapes)		Places Scenes Shape Pictures Memory Number		Party Time 3 Memory Number 2		Snapshots Memory Number		Family Letter		Counting Jar Numerals Every Day Shape Walk

Building Blocks • *Assessment*

Weekly Record Sheet

Directions:
- Under the week number, write the date you started that week's activities.
- In the ✔ columns, put a ✔ when an activity is completed.
- For Whole Group and Math Throughout the Year, put a ✔ for each time the class does the activity.
- For the other columns, put a ✔ when you start each activity.

Week	✔	Whole Group	✔	Hands On Math Center	✔	Computer Center	✔	Small Group	✔	Home Connection	✔	Math Throughout the Year	✔
19 Date ____		Count and Move in Patterns X-Ray Vision 1 Mr. Mixup (Counting) Knock It Down Mr. Mixup (Comparing)		Places Scenes X-Ray Vision 1 Dinosaur Shop (Fill Orders)		Dinosaur Shop 2 Pizza Pizzazz 3		X-Ray Vision 1 Dinosaur Shop (Fill Orders)		Family Letter		Counting Wand (Counting Up) Dinosaur Shop (Dramatic Play)	
20 Date ____		Count and Move (Forward and Back) As Long As My Arm Listen and Copy Snapshots How Many Now? (Hidden version)		As Long As My Arm X-Ray Vision 1 Compare Capacities		Deep Sea Compare		How Many Now? (Hidden version) X-Ray Vision 1		Family Letter		Compare Lengths Compare Weights Line Up by Height	

10

Building Blocks • Assessment

Weekly Record Sheet

Directions
- Under the week number, write the date you started that week's activities.
- In the ✔ columns, put a ✔ when an activity is completed.
- For Whole Group and Math Throughout the Year, put a ✔ for each time the class does the activity.
- For the other columns, put a ✔ when you start each activity.

Week	✔	Whole Group	✔	Hands On Math Center	✔	Computer Center	✔	Small Group	✔	Home Connection	✔	Math Throughout the Year	✔
21 _____ Date		Count and Move in Patterns What's the Missing Step? "Ten Little Monkeys" Number Snapshots What's This Step?		X-Ray Vision 1 Compare Capacities What's the Missing Step?		Build Stairs 3 Workin' on the Railroad		What's the Missing Step? Length Riddles		Family Letter		Line Up by Height Sense of Time	
22 _____ Date		Measure Length X-Ray Vision 2 Blast Off Mr. Mixup's Measuring Mess "Ten Little Monkeys" I'm Thinking of a Number (Length)		Places Scenes Compare Game Road Blocks X-Ray Vision 2		Reptile Ruler Number Compare 1		Length Riddles X-Ray Vision 2		Family Letter		How Many Seconds? Measure Capacities Weigh Blocks	

Building Blocks • Assessment

Weekly Record Sheet

Directions:
- Under the week number, write the date you started that week's activities.
- In the ✔ columns, put a ✔ when an activity is completed.
- For Whole Group and Math Throughout the Year, put a ✔ for each time the class does the activity.
- For the other columns, put a ✔ when you start each activity.

Week	✔	Whole Group	✔	Hands On Math Center	✔	Computer Center	✔	Small Group	✔	Home Connection	✔	Math Throughout the Year
23 Date		Blast Off Puzzles I'm Thinking of a Number I Spy Guess My Rule		Pattern Block Puzzles X-Ray Vision 2 Pattern Block Cutouts		Piece Puzzler 1 Piece Puzzler 2		What's the Missing Card? Pattern Block Puzzles		Family Letter		Counting Jar How Many Seconds? Line Up—Who's First? Measure Capacities
24 Date		Finger Word Problems Places Scenes (Adding) "Five Little Monkeys" Snapshots (Adding)		Places Scenes (Adding) Dinosaur Shop (Adding) Tangram Puzzles Tangram Pictures		Pizza Pizzazz 4 Dinosaur Shop 3		How Many Now? Dinosaur Shop (Adding)		Family Letter		Counting Jar Line Up—Who's First? Snack Time

12

Building Blocks • *Assessment*

Copyright © The McGraw-Hill Companies, Inc. Permission is granted to reproduce for classroom use.

Weekly Record Sheet

Directions:
- Under the week number, write the date you started that week's activities.
- In the ✔ columns, put a ✔ when an activity is completed.
- For Whole Group and Math Throughout the Year, put a ✔ for each time the class does the activity.
- For the other columns, put a ✔ when you start each activity.

Week	✔	Whole Group	✔	Hands On Math Center	✔	Computer Center	✔	Small Group	✔	Home Connection	✔	Math Throughout the Year	✔
25 Date ___		Finger Word Problems I'm Thinking of a Number (Length) Count and Move (Forward and Back) X-Ray Vision 2 I'm Thinking of a Number (Ruler) I'm Thinking of a Number (Clues)		Places Scenes (Adding) Dinosaur Shop (Adding) Tangram Puzzles Adding: Board Game X-Ray Vision 2		Dinosaur Shop 3 Off the Tree		X-Ray Vision 2 Snapshots (Adding)		Family Letter		Counting Jar Line Up—Who's First? Snack Time	
26 Date ___		Finger Word Problems Dinosaur Shop (Make It Right) Line Up—Who's First? "Five Little Ducks" Snapshots (Adding) Gone Fishing Count and Move (Forward and Back)		Dinosaur Shop (Adding) Pizza Game 2 Adding: Board Game X-Ray Vision 2		Dinosaur Shop 4 Ordinal Construction Company Countdown Crazy		Dinosaur Shop (Make It Right) Pizza Game 2		Family Letter		Counting Jar Snack Time	

Building Blocks • Assessment

13

Weekly Record Sheet

Directions:
- Under the week number, write the date you started that week's activities.
- In the ✔ columns, put a ✔ when an activity is completed.
- For Whole Group and Math Throughout the Year, put a ✔ for each time the class does the activity.
- For the other columns, put a ✔ when you start each activity.

Week	✔	Whole Group	✔	Hands On Math Center	✔	Computer Center	✔	Small Group	✔	Home Connection	✔	Math Throughout the Year
27 Date ___		Shape Parts (Triangles) Shape Step (Properties) **Building Shapes** Feely Box (Describe) Snapshots (Shapes) I Spy (Properties)		Pattern Block Puzzles Tangram Puzzles Pattern Block Cutouts		Shape Parts 1 Piece Puzzler 2		Building Shapes Feely Box (Describe)		Family Letter		Counting Jar I Spy (Properties) Line Up—Who's First? Tape Shapes
28 Date ___		Mr. Mixup (Shapes) Shape Step (Properties) Feely Box (Describe) Snapshots (Shape Parts)		Shape Transparencies Pattern Block Puzzles Tangram Puzzles Pattern Block Cutouts Building Shapes		Shape Parts 2 Piece Puzzler 3		Building Shapes Shape Step (Properties)		Family Letter		Counting Jar I Spy (Properties) Tape Shapes

14

Building Blocks • *Assessment*

Small Group Record Sheets

The Small Group Record Sheets on the following pages include opportunities to record observations of children as they participate in Small Group activities.

Procedure:

1. Find the correct Small Group Record Sheet for the activity being conducted; they are labeled for each week.

2. Copy each Record Sheet and fill in children's names.

3. Communicate to children what you will be assessing and what behaviors you will be looking for.

4. Conduct the activity, recording children's behaviors during or after the lesson.

5. Review findings. The learning trajectory levels are provided on the sheet to guide your review.

6. Provide feedback to children and their families about children's achievements based on your informal assessments.

Small Group Record Sheet
Week 1 Activity—Explore Manipulatives

5 Corresponder
6 Counter (Small Numbers)
7 Producer (Small Numbers)

Child's name:	Observations/Trajectory Level:	Comments:

Building Blocks • *Assessment*

Small Group Record Sheet
Week 2 Activity—Find and Make Groups

1 Small Collection Namer
3 Maker of Small Collections
4 Perceptual Subitizer to 4
5 Perceptual Subitizer to 5

Child's name:	Finds groups to:	Strategies/Trajectory Level:	Comments:

Building Blocks • *Assessment*

Small Group Record Sheet
Week 3 Activity—Demonstrate Counting

5 Corresponder
6 Counter (Small Numbers)
7 Producer (Small Numbers)

Child's name:	Names numbers:	Strategies/Trajectory Level:	Comments:

Building Blocks • *Assessment*

Small Group Record Sheet
Week 3 Activity—Make Number Pizzas

6 Counter (Small Numbers)
7 Producer (Small Numbers)
8 Counter (10)

Child's name:	Numbers made well:	Strategies/Trajectory Level:	Comments:

Small Group Record Sheet

Week 4 Activity—Match and Name Shapes

2 **Shape Recognizer, Typical**
4 **Shape Recognizer, Circles, Squares, and Triangles+**
6 Shape Recognizer, All Rectangles

Child's name:	Shapes known and Trajectory Level:	Child's explanation(s):	Comments:

Small Group Record Sheet
Week 5 Activity—Straw Shapes

> **5** **Constructor of Shapes from Parts, Looks Like**
> **13** Constructor of Shapes from Parts, Exact

Child's name:	Shapes made:	Accuracy/Trajectory Level:	Comments:

Small Group Record Sheet
Week 5 Activity—Is It or Not?

6 Shape Recognizer,
All Rectangles
**9 Shape Recognizer,
More Shapes**
10 Shape Identifier

Child's name:	Shapes known:	Justification given/ Trajectory Level:	Comments:

Small Group Record Sheet
Week 6 Activity—Pizza Game 1

5 Corresponder
6 Counter (Small Numbers)
7 Producer (Small Numbers)

Child's name:	Counts, Produces, Compares (Trajectory Level):	Strategies/Trajectory Level:	Comments:

Small Group Record Sheet
Week 6 Activity—Make Number Pizzas

6 Counter (Small Numbers)
7 Producer (Small Numbers)
8 Counter (10)

Child's name:	Numbers made well:	Strategies/Trajectory Level:	Comments:

Building Blocks • *Assessment*

Small Group Record Sheet
Week 7 Activity—Get Just Enough

8 Counting Comparer (Same Size)
9 Counting Comparer (5)
11 Counting Comparer (10)

Child's name:	Accurately groups up to:	Strategies used (matching, counting, and so on)/ Trajectory Level:	Comments:

Small Group Record Sheet
Week 7 Activity—Compare Game

> 6 Matching Comparer
> 8 Counting Comparer (Same Size)
> 9 Counting Comparer (5)
> 11 Counting Comparer (10)

Child's name:	Counts well:	Compares numbers:	Strategies/ Trajectory Level:	Comments:

Building Blocks • *Assessment*

Small Group Record Sheet

Week 8 Activity—Number Jump

Child's name:	Accurately "produces" a number of jumps up to:	Strategies/Trajectory Level:	Comments:

Small Group Record Sheet
Week 8 Activity—Number Race

Child's name:	Counts dots on cube to:	Counts spaces on board accurately:	Strategies/ Trajectory Level:	Comments:

Building Blocks • *Assessment*

Small Group Record Sheet
Week 9 Activity—Is It or Not?

9 Shape Recognizer, More Shapes
10 Shape Identifier
12 Parts of Shapes Identifier

Child's name:	Shapes/Trajectory Level:	Child's justification(s):	Comments:

Small Group Record Sheet
Week 9 Activity—Memory Geometry

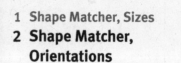

1 Shape Matcher, Sizes
2 **Shape Matcher, Orientations**
3 Shape Matcher, More Shapes

Child's name:	Matches shapes:	Names shapes:	Remembers array:	Strategies/ Trajectory Level:	Comments:

Building Blocks • *Assessment*

Small Group Record Sheet
Week 10 Activity—Guess My Rule

7 **Side Recognizer**
9 **Shape Recognizer,
More Shapes**
10 Shape Identifier
12 **Parts of Shapes Identifier**

Child's name:	Guesses which rules/ Trajectory Level	Reasoning displayed:	Comments:

Building Blocks • *Assessment*

Small Group Record Sheet
Week 10 Activity—Shape Step

6 Shape Recognizer,
  All Rectangles
7 Side Recognizer
**9 Shape Recognizer,
  More Shapes**
10 Shape Identifier

Child's name:	Finds which shapes?	Strategies/Trajectory Level:	Comments:

Building Blocks • *Assessment*

Small Group Record Sheet
Week 11 Activity—How Many Now?

2 Nonverbal +/−
3 Small Number +/−
4 Find Results +/−

Child's name:	Can "add on" 1, 2...?	Strategies/Trajectory Level:	Comments:

Small Group Record Sheet
Week 11 Activity—Number Choice

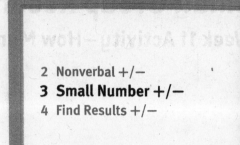

2 Nonverbal +/−
3 Small Number +/−
4 Find Results +/−

Child's name:	Chooses number for best results?	Strategies/Trajectory Level:	Comments:

Building Blocks • *Assessment*

Small Group Record Sheet
Week 12 Activity—Mr. Mixup

8 Counter (10)
9 Counter and Producer (10+)
10 Counter Backward from 10

Child's name:	Description of mistakes/ Trajectory Level:	Reasoning displayed:	Comments:

Small Group Record Sheet
Week 12 Activity—Dinosaur Shop
(Sort and Label)

8 Counter (10)

9 Counter and Producer (10+)

10 Counter Backward from 10

Child's name:	Sorts according to given criteria? Counts dinosaurs to?	Strategies/Trajectory Level:	Comments:

Building Blocks • *Assessment*

Small Group Record Sheet
Week 13 Activity—Build Cube Stairs

Child's name:	Builds stairs? Extends?	Strategies/Trajectory Level:	Comments (Does child understand patterns? Cores?):

Small Group Record Sheet

Week 13 Activity—How Many Now?

Child's name:	Can "add on" 1, 2...?	Strategies/Trajectory Level:	Comments:

Building Blocks • *Assessment*

Small Group Record Sheet

Week 14 Activity—Feely Box (Match and Name)

6 Shape Recognizer, All Rectangles
7 Side Recognizer
9 Shape Recognizer, More Shapes
10 Shape Identifier
12 Parts of Shapes Identifier

Child's name:	Shapes identified:	Justifications given/ Trajectory Level:	Comments:

Small Group Record Sheet
Week 15 Activity—Guess My Rule

7 Side Recognizer
**9 Shape Recognizer,
    More Shapes**
10 Shapes Identifier
12 Parts of Shapes Identifier
14 Shapes Class Identifier

Child's name:	Guesses which rules/ Trajectory Level	Reasoning displayed:	Comments:

Building Blocks • *Assessment*

Small Group Record Sheet
Week 15 Activity—Shape Step

	6 Shape Recognizer, All Rectangles
	7 **Side Recognizer**
	9 **Shape Recognizer, More Shapes**
	10 Shape Identifier

Child's name:	Finds which shapes?	Strategies/Trajectory Level:	Comments:

Building Blocks • *Assessment*

Small Group Record Sheet
Week 16 Activity—Pattern Strips

Child's name:	Matches, really duplicates, extends?	Strategies/Trajectory Level:	Comments (Does child understand patterns? Cores?):

Building Blocks • *Assessment*

Small Group Record Sheet

Week 17 Activity—Pattern Strips (The Core)

Child's name:	Duplicates cores? Extends?	Strategies/Trajectory Level:	Comments (Does child understand patterns? Cores?):

Small Group Record Sheet
Week 18 Activity—Snapshots

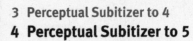

3 Perceptual Subitizer to 4
4 Perceptual Subitizer to 5
5 Conceptual Subitizer to 5+

Child's name:	Subitizes to/ Trajectory Level:	Child's explanation(s):	Comments:

Building Blocks • *Assessment*

Small Group Record Sheet

Week 18 Activity—Memory Number

7 Producer (Small Numbers)

8 Counter (10)

9 Counter and Producer (10+)

10 Counter Backward from 10

Child's name:	Matches amounts:	Names numbers:	Remembers array:	Strategies/ Trajectory Level:	Comments:

Small Group Record Sheet
Week 19 Activity—X-Ray Vision 1

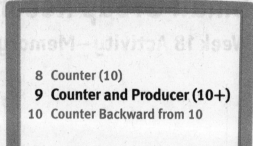

8 Counter (10)
9 Counter and Producer (10+)
10 Counter Backward from 10

Child's name:	Counts correctly?	Strategies/Trajectory Level:	Comments:

Small Group Record Sheet

Week 19 Activity—Dinosaur Shop (Fill Orders)

8 Counter (10)
9 Counter and Producer (10+)
10 Counter Backward from 10

Child's name:	Produces up to 10? Keeps track of unordered items?	Strategies/Trajectory Level:	Comments:

Small Group Record Sheet

Week 20 Activity—How Many Now?

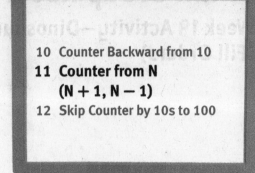

10 Counter Backward from 10
**11 Counter from N
(N + 1, N − 1)**
12 Skip Counter by 10s to 100

Child's name:	Can "add on" 1, 2...?	Strategies/Trajectory Level:	Comments:

Building Blocks • *Assessment*

Small Group Record Sheet
Week 20 Activity—X-Ray Vision 1

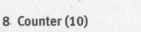

8. Counter (10)
9 Counter and Producer (10+)
10 Counter Backward from 10

Child's name:	Counts correctly?	Strategies/Trajectory Level:	Comments:

Small Group Record Sheet
Week 21 Activity—What's the Missing Step?

10 Counter Backward from 10
**11 Counter from N
(N + 1, N − 1)**
12 Skip Counter by 10s to 100

Child's name:	Identifies missing step?	Strategies/Trajectory Level:	Comments:

Building Blocks • *Assessment*

Small Group Record Sheet
Week 21 Activity—Length Riddles

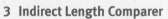

3 Indirect Length Comparer
5 End-to-End Length Measurer
7 Length Unit Relator

Child's name:	Can measure and find objects?	Strategies/Trajectory Level:	Comments:

Small Group Record Sheet
Week 22 Activity—Length Riddles

3 Indirect Length Comparer
5 End-to-End Length Measurer
6 Length Unit Iterator

Child's name:	Can measure and find objects?	Strategies/Trajectory Level:	Comments:

Building Blocks • *Assessment*

Small Group Record Sheet
Week 22 Activity—X-Ray Vision 2

Child's name:	Counts correctly?	Strategies/Trajectory Level:	Comments:

Small Group Record Sheet

Week 23 Activity—What's the Missing Card?

Child's name:	Identifies missing card?	Strategies/Trajectory Level:	Comments:

Building Blocks • *Assessment*

Small Group Record Sheet
Week 23 Activity—Pattern Block Puzzles

Child's name:	Places shapes accurately with no gaps? Uses trial and error and/or imagery?	Strategies/Trajectory Level:	Comments:

Small Group Record Sheet
Week 24 Activity—How Many Now?

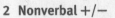

2 Nonverbal +/−
3 Small Number +/−
4 Find Result +/−

Child's name:	Can "add on" 1, 2...?	Strategies/Trajectory Level:	Comments:

Building Blocks • *Assessment*

Small Group Record Sheet
Week 24 Activity—Dinosaur Shop (Adding)

2 Nonverbal +/−
3 Small Number +/−
4 Find Result +/−

Child's name:	Finds totals? Can child count up to, or produce, 10 or higher with dinosaurs and money? Check whether child can count unordered sets of dinosaurs.	Strategies (counts all, counts on, subitizes, composes, and so on)/ Trajectory Level:	Comments:

Small Group Record Sheet
Week 25 Activity—X-Ray Vision 2

Child's name:	Counts correctly?	Strategies/Trajectory Level:	Comments:

Building Blocks • *Assessment*

Small Group Record Sheet

Week 25 Activity—Snapshots (Adding)

4 Perceptual Subitizer to 5
5 Conceptual Subitizer to 5+
6 Conceptual Subitizer to 10

Child's name:	Subitizes to/ Trajectory Level:	Child's explanation(s):	Comments:

Small Group Record Sheet

Week 26 Activity—Dinosaur Shop
(Make It Right)

4 Find Result +/—
5 **Find Change +/—**
6 **Make it N +/—**
7 Counting Strategies +/—

Child's name:	Correctly adjusts total?	Strategies (Adds on? Counts from 1? Trial and error?)/Trajectory Level:	Comments (Does child understand? Seems to know what number to add on?):

Building Blocks • *Assessment*

Small Group Record Sheet
Week 26 Activity—Pizza Game 2

10 Counter Backward from 10
11 **Counter from N**
    (N + 1, N − 1)
12 Skip Counter by 10s to 100

Child's name:	Accurately carries out each part of the activity?	Strategies (Does child quickly know how to make the target number? Counts up from an existing number? Uses trial and error?)/Trajectory Level:	Comments:

Small Group Record Sheet
Week 27 Activity—Building Shapes

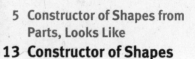

5 Constructor of Shapes from Parts, Looks Like

13 Constructor of Shapes from Parts, Exact

Child's name:	Builds squares, rectangles, triangles?	Accuracy/Trajectory Level:	Comments (Understands and discusses shapes' attributes?):

Small Group Record Sheet
Week 27 Activity—Feely Box (Shapes)

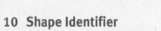

10 Shape Identifier
12 Parts of Shapes Identifier
14 Shape Class Identifier

Child's name:	Correct with which shapes? Describes shapes completely?	Justification given/ Trajectory Level:	Comments (Does child understand and communicate about the shapes?):

Building Blocks • *Assessment*

Small Group Record Sheet
Week 28 Activity—Building Shapes

> 5 Constructor of Shapes from Parts, Looks Like
> 13 **Constructor of Shapes from Parts, Exact**

Child's name:	Builds rhombus, triangle, other?	Accuracy/Trajectory Level:	Comments (Understands and discusses shapes' attributes?):

Building Blocks • *Assessment*

Small Group Record Sheet
Week 28 Activity—Shape Step (Properties)

Child's name:	What properties does child correctly identify?	Justification given/ Trajectory Level:	Comments (Understands and communicates about shapes and their attributes?):

Building Blocks • *Assessment*

Small Group Record Sheet
Week 29 Activity—Compare Game (Adding)

3 Small Number +/—
4 Find Result +/—
5. Find Change +/—

Child's name:	Finds sums? Can tell each number of the sum?	Strategies (counts all, counts on, subitizes, composes, and so on)/ Trajectory Level:	Comments:

Building Blocks • *Assessment*

Small Group Record Sheet
Week 29 Activity—Pattern Block Puzzles

Child's name:	Places shapes accurately with no gaps? Uses trial and error or imagery?	Strategies/Trajectory Level:	Comments:

Building Blocks • *Assessment*

Small Group Record Sheet
Week 30 Activity—Compare Game (Adding)

3 Small Number +/−
4 Find Result +/−
5 Find Change +/−

Child's name:	Finds sums? Can tell each number of the sum?	Strategies (counts all, counts on, subitizes, composes, and so on)/ Trajectory Level:	Comments:

**Building Blocks • ** *Assessment*

Small Group Record Sheet
Week 30 Activity—Places Scenes (Parts)

Child's name:	Counts accurately? Finds sums?	Strategies/Trajectory Level:	Comments:

Learning Trajectory Records

The Learning Trajectory Records on the following pages provide opportunities to identify children's levels on the Developmental Progressions from each learning trajectory or topic. Once a child exhibits a certain level of thinking for a trajectory, teachers should begin to encourage work towards the next level.

Such assessments are intended to inform instruction by helping teachers identify their children's proficiency in each area of mathematics. The results can also be compared to the *Building Blocks* software reports to check reliability.

Procedure:

1. Collect any materials needed for each assessment.

2. Assess children individually.

3. Make an educated guess as to what level the child is capable of exhibiting or start with the first question.

4. Ask the first question. Provide time for the child to respond.

5. Evaluate and record the response.

6. If the child responds correctly with confidence, move on to the next question.

7. If the child responds incorrectly or hesitates, provide a prompt to make sure that the child has not misunderstood.

8. If the child again responds incorrectly or is confused, move backward to an earlier level, if one exists, to see what level of thinking the child is capable of showing.

9. Record the date for each level that was exhibited on the assessment on the Trajectory Progress Chart.

Child's Name _____

Learning Trajectory Record: Counting

Materials: 15 cubes (wooden or connecting), or other small manipulatives that children would enjoy counting; a piece of paper

Age	Descriptors	Questions	Date	Response
1	**Pre-Counter.** No verbal counting. Names some number words with no sequence.	*Question 1.* For Pre-Counter through to Reciter (10)		
	Chanter. Chants "sing-song" or sometimes indistinguishable number words ("onetwothree").	Ask: **How high can you count? Start at 1 and show me.** If a child stops, you can prompt to continue once per stop, but at as many different points in counting as necessary; for example, **"What comes next? Can you go higher?"**		
2	**Reciter.** Verbally counts with separate words, not necessarily in the correct order ("one, two, three, four, six, seven").			
3	**Reciter (10).** Verbally counts to 10 or beyond.			

Building Blocks • *Assessment*

73

Learning Trajectory Record: Counting

Age	Descriptors	Questions	Date	Response
3	**Corresponder.** Keeps one-to-one correspondence between counting words and objects (one word for each object), at least for small groups of objects laid in a line ("1, 2, 3, 4, 5"). May answer the question "how many?" by re-counting the objects.	*Question 2.* For Corresponder through to Counter (Small Numbers) Place five cubes in a line. Say: **Count these blocks to tell me how many there are.** If the child repeated the last counting word or otherwise showed she understood that "5" was the number of cubes (for example, "1, 2, 3, 4, 5: 5 blocks!"), then they have achieved this level. Otherwise, ask: **How many blocks are there?** The child must say *five* without needing to recount the cubes.		
4	**Counter (Small Numbers).** Accurately counts objects in a line to 5 and answers the question "how many?" with the last number counted. When objects are visible, and especially with small numbers, begins to understand cardinality ("1, 2, 3, 4, 5... *five!*").			
	Producer (Small Numbers). Counts out objects to 5.	*Question 3.* Place about ten cubes near the paper. Say: **Please help me put four blocks on your paper.** If the child counts past 4, remind that the goal is four and allow her or him to try again one time.		

Child's Name _____

Learning Trajectory Record: Counting

Age	Descriptors	Questions	Date	Response
4	**Counter (10).** Counts structured arrangements of objects to 10. Accurately counts a line of nine blocks and says there are 9. May be able to tell the number just before or just after another number, but only by counting up from 1. Ask, **What comes after 4? "1, 2, 3, 4, 5. 5!"**	*Question 4.* Show the child eight cubes laid in a line. Say: **Show me how you can count the blocks, and tell me how many there are** (child is allowed to move the cubes). Cover the cubes with a piece of paper and immediately ask: **Now, I am covering them all! How many blocks are under here?** (Child is not allowed to recount.)		
5	**Counter and Producer—Counter To (10+).** Counts and counts out objects accurately to 10, then beyond, up to 30. Has explicit understanding of cardinality (that numbers tell how many). Keeps track of objects that have and have not been counted, even in different arrangements. Gives next number to 20s or 30s.	*Question 5.* Lay fifteen cubes in an unordered, scrambled arrangement. Say: **Show me how you can count the blocks and tell me how many there are.** (Child is allowed to move the cubes.) Leave out the cubes. *Question 6.* Place the shopping cart in front of the child and about 15 cubes near the paper. Say: **Put exactly 10 blocks on your paper.**		
	Counter Backward from 10. Counts backward from 10 to 1, verbally, or when removing objects from a group. "10, 9, 8, 7, 6, 5, 4, 3, 2, 1!"	*Question 7.* Say: **Help me count backward from 10 to 0 to blast off: 10, 9...** If no response follows, say: **What comes next?** You can say "keep going" to encourage the child to keep counting, if they stop.		

Learning Trajectory Record: Counting

Age	Descriptors	Questions	Date	Response
6	**Counter from N (N+1, N−1).** Counts verbally and with objects from numbers other than 1. Can determine numbers just after or just before immediately. Asked, **"What comes just before 7?" says, "Six!"**	Remove all objects. *Question 8.* Say: **Let's practice some counting. Please count to 10, starting at 4.** (If a child starts at 1, interrupt by saying: "Please start at 4 and count to 10.") *Question 9.* **Can you help me, what number comes right after 5?** Child is at this level if they answer immediately, without counting from 1.		

Building Blocks • *Assessment*

Child's Name

Learning Trajectory Record: Comparing and Ordering Number

Materials: Cubes (or other small manipulatives that children would enjoy counting) in two colors, small toys

Age	Descriptors	Questions	Date	Response
2	**Perceptual Comparer.** Compares collections that are quite different in size (for example, one is at least twice the other). If the collections are similar, can compare very small collections.	*Question 1.* Show the child two cubes of one color and four cubes of another color. Ask: **Which pile has more?**		
2–3	**First-Second Ordinal Counter.** Identifies the "first" and often "second" objects in a sequence.	*Question 2.* Place four toys lining up at a toy house or other object. Say: **They are waiting in line. Which one is first in line? Which one is second in line?**		
3	**Nonverbal Comparer of Similar Items.** (1–4 items) Compares collections of 1–4 items verbally or nonverbally ("just by looking"). The items must be the same.	*Question 3.* Show three cubes arranged like this ••• and three cubes arranged like this •••• and say: **What do you think? Do these two groups have the same number?**		
	Nonverbal Comparer of Dissimilar Items. Matches small, equal collections of shells and dots, thus showing that they are the same number.	*Question 4.* Show large (wooden inch) cubes arranged like this •••• and small counters arranged like this ▢▢▢▢ and say: **Do these two groups have the same number?**		

Learning Trajectory Record: Comparing and Ordering Number

Age	Descriptors	Questions	Date	Response
4	**Matching Comparer.** Compares groups of 1–6 by matching.	*Question 5.* From Matching Comparer through to Counting Comparer (Same Size)		
	Counting Comparer (Same Size). Accurately compares via counting, but only when objects are about the same size and groups are small.	Show the child five cubes of one color and six cubes of another color. Ask: **Are there the same numbers of cubes in each group, or does one group have more?**		

Child's Name _____

Learning Trajectory Record: Comparing and Ordering Number

Age	Descriptors	Questions	Date	Response
5	**Counting Comparer (5).** Compares with counting, even when larger collection's objects are smaller. Figures out how many more or less. Accurately counts two equal collections, and says they have the same number, even if one collection has larger blocks.	*Question 6.* Put out a pile of four large (wooden inch) cubes and five small counters. Ask: **Are there more blocks or more counters or is there the same number?**		
	Ordinal Counter. Identifies and uses ordinal numbers from *first* to *tenth*. Can identify who is "third in line."	*Question 7.* Place ten small toys lining up at a toy house or other object. Say: **They are waiting in line. Which one is third in line?** Repeat with other ordinal numbers, in random order, including fourth to tenth.		

Building Blocks • *Assessment*

Learning Trajectory Record: Comparing and Ordering Number

Age	Descriptors	Questions	Date	Response
6	**Counting Comparer (10).** Compares with counting, even when larger collection's objects are smaller, up to 10. Accurately counts two collections of 9 each, and says they have the same number, even if one collection has larger blocks.	*Question 8.* Show the child nine blue cubes and eleven red cubes. Ask: **Are there more blue cubes or red cubes or is there the same number?**		
	Mental Number Line to 10. Uses internal images and knowledge of number relationships to determine relative size and position. Which number is closer to 6:4 or 9?	Have no objects available. *Question 9.* **Which number is closer to 7: 3 or 4?** *Question 10.* **What number is between 5 and 7?**		

Building Blocks • *Assessment*

Child's Name _____

Learning Trajectory Record: Numerals

Materials: paper and pencil; box; 6 objects; numeral cards 0-5; dot cards 0-5

Age	Descriptors	Questions	Date	Response
3	**Quantity Representer.** Represents and recalls sets with pictographic, iconic representations of quantity.	Ask student to put three objects into a box and close the box. [b] Say: **"Can you write or draw on this paper so we remember how many are in the box?"**		
4	**Numeral Representer.** Matches small sets (1–5) with the corresponding numbers. Represents and recalls the size of sets using those numbers.	Show numeral cards (1-5) and the dot cards (1-5) . [b] Say: **"Match the numbers with the dot cards."**		
4-5	**Functional Numeral User.** Uses numerals to represent and communicate quantity. For example, uses numerals to remember results of counting or to compare quantities.	Work with the student to put one object under a sheet of paper (or a container), two under another, and so forth up to six. [b] Say: **"What could you write on each so we remember how many there are?"** After the student is finished, say: **"Which one has four? Which one has the most?"**		

Learning Trajectory Record: Recognizing Number and Subitizing

Materials: Cubes (or other small manipulatives that children would enjoy counting) in two colors

Age	Descriptors	Questions	Date	Response
2	**Small Collection Namer.** Names groups of 1 to 2, sometimes 3.	*Question 1.* Show the child two cubes. Ask: **How many blocks are there here?**		
3	**Maker of Small Collections.** Nonverbally makes a small collection (no more than 4, usually 1–3) with the same number as another collection. May be able to name the number.	*Question 2.* Place six cubes near the child. Put out three cubes in front of you for two seconds, and then hide them. Say: **Can you make the same number of blocks?**		

Child's Name

Learning Trajectory Record: Recognizing Number and Subitizing

Age	Descriptors	Questions	Date	Response
4	**Perceptual Subitizer to 4.** Instantly recognizes collections up to 4 briefly shown and verbally names the number of items.	*Question 3.* Show four cubes to the child for two seconds, and then remove them and ask: **How many?**		

Building Blocks • *Assessment*

Learning Trajectory Record: Recognizing Number and Subitizing

Age	Descriptors	Questions	Date	Response
5	**Perceptual Subitizer to 5.** Instantly recognizes briefly shown collections up to 5 and verbally names the number of items. Shown 5 objects briefly, says "5."	*Question 4.* Show five cubes to the child for two seconds, and then remove them and ask: **How many?**		
	Conceptual Subitizer to 5+. Verbally labels all arrangements to about 5, when shown only briefly.			
	Conceptual Subitizer to 10. Verbally labels most briefly shown arrangements to 6, then up to 10, using groups. "In my mind, I made two groups of 3 and one more, so 7."	*Question 5.* Show ten cubes to the child, seven in one hand and three in the other, for two seconds, and then remove them and ask: **How many?**		

Child's Name

Learning Trajectory Record: Composing Number

Materials: Cubes (or other small manipulatives that children would enjoy counting) in two colors

Age	Descriptors	Questions	Date	Response
5	**Composer to 4, then 5.** Knows number combinations. Quickly names parts of any whole, or the whole given the parts.	*Question 1.* From Composer to 4, then 5 through to Composer to 7 Lay ten cubes on the table in a scrambled arrangement. Say: **I put four blocks here. Count with me. 1, 2, 3, 4. 4!** (Lay them in a straight line as you count.) **Now, I'm going to hide some.** (Cover the cubes with a cloth; then hide three; then remove the cover to show the remaining two.) Ask: **How many am I hiding?** Repeat with five and seven boxes.		
6	**Composer to 7.** Knows number combinations to totals of 7. Quickly names parts of any whole, or the whole given parts. Doubles to 10.			

Learning Trajectory Record: Adding and Subtracting

Materials: Cubes (or other small manipulatives that children would enjoy counting) in two colors

Age	Descriptors	Questions	Date	Response
3	**Nonverbal +/−.** Adds and subtracts very small collections nonverbally.	*Question 1.* Show two cubes, and then cover them with a cloth. Tell children to watch carefully, and then put one more cube under the cloth. Give the child six cubes and ask, **Make a group of blocks look just like I have** (gesture to the cloth).		

Child's Name

Learning Trajectory Record: Adding and Subtracting

Age	Descriptors	Questions	Date	Response
4	**Small Number +/−.** Finds sums for joining problems up to 3 + 2 by counting all with objects.	*Question 2.* You have two balls and get one more. How many do you have in all?		

Learning Trajectory Record: Adding and Subtracting

Age	Descriptors	Questions	Date	Response
5	**Find Result +/−.** Finds sums for joining and part-part-whole problems by direct modeling, counting all, with objects. Solves take-away problems by separating with objects.	*Question 3.* You have two red balls and three blue balls. How many in all? *Question 4.* You have five balls and give two to Tom. How many do you have left?		
	Find Change +/−. Finds the missing addend (5 + ____ = 7) by adding on objects. Compares by matching in simple situations.	*Question 5.* You have five balls and then get some more. Now you have seven in all. How many did you get? *Question 6.* Here are six dogs and four balls. If we give a ball to each dog, how many dogs won't get a ball?		
	Make It N +/−. Adds on objects to make one number into another, without counting from "one." Asked, "This puppet has four balls, but she should have six. Make it six," puts up four fingers on one hand, immediately counts up from four while putting up two fingers on the other hand, saying, "5, 6"; and then counts or recognizes the two fingers.	*Question 7.* You have five apples but want to have seven. How do you make it seven?		

Copyright © The McGraw-Hill Companies, Inc. Permission is granted to reproduce for classroom use.

Building Blocks • Assessment

Child's Name _____

Learning Trajectory Record: Adding and Subtracting

Age	Descriptors	Questions	Date	Response
6	**Counting Strategies +/−.** Finds sums for joining (you had eight apples and get three more...) and part-part-whole (six girls and five boys...) problems with finger patterns or by adding on objects or counting on. "How much is 4 and 3 more?" "Fourrrr... five, six, seven (uses rhythmic or finger pattern). Seven!" Solves missing addend ($3 + \_\_\_ = 7$) or compare problems by counting up.	Have no objects available. Allow, but do not mention or encourage, use of fingers. *Question 8.* **How much is 6 + 4?** *Question 9.* **You have seven apples. How many more would you need to have 9?** *Question 10.* **How much is 5 and 4 more?**		

Building Blocks • *Assessment*

Learning Trajectory Record: Length Measuring

Materials: Two pencils of different lengths, a piece of string, Connecting Cubes, two books of different sizes, ruler

Age	Descriptors	Questions	Date	Response
4	**Length Direct Comparer.** Physically aligns two objects to determine which is longer or whether they are the same length. Stands two sticks next to each other on a table and says, "This one's bigger."	*Question 1.* Give the child two pencils of different lengths. Ask: **Which of these is longer?**		
5	**Indirect Length Comparer.** Compares the length of two objects by representing them with a third object. Compares length of two objects with a piece of string.	*Question 2.* Give the child two different pencils of different lengths, laid several feet apart, and a piece of string. Ask: **Which of these pencils is longer? Tell me without moving the pencils. You can use this string to help you.**		
6	**Serial Orderer to 6+.** Orders lengths marked into units (1–6, then beyond). Given towers of cubes, puts in order, 1 to 6.	*Question 3.* Give the child connecting cube towers from 3 to 10 in random order. (Do not allow the child to take them apart or connect them.) Ask: **Put these in order from smallest to largest. Line them up so their bottoms are at the edge of the table.**		
	End-to-End Length Measurer. Lays or connects units end-to-end to measure.	*Question 4.* Give the child a book and a pile of connecting cubes. Ask: **Would you please measure how long this book is with these cubes?**		
7	**Length Unit Relater and Repeater.** Iterates a single unit to measure. Sees need for identical units. Uses rulers with guidance. Measures a book's length accurately with a ruler.	*Question 5.* Give the child a different size book and a ruler. Ask: **Would you please measure how long this book is with this ruler?**		

Building Blocks • *Assessment*

Child's Name _____

Learning Trajectory Record: Geometry (Shapes)

Materials: Shape Sets

Age	Descriptors	Questions	Date	Response
2	**Shape Matcher—Identical.** Matches familiar shapes (circle, square, typical triangle) with same size and orientation. —**Sizes.** Matches familiar shapes with different sizes. —**Orientations.** Matches familiar shapes with different orientations.	*Question 1.* Use Shape Sets for these questions. —**Identical** Display shape C from the blue Shape Set. Under it display shapes B, T, C, and N from the yellow Shape Set. Ask: **Which yellow shape** (gesture) **matches this blue shape** (gesture)**?** (Answer: C) —**Sizes** Display shape G from the blue Shape Set. Under it display shapes E, I, Z7, and Z1 from the yellow Shape Set. Ask: **Which yellow shape** (gesture) **is the same shape as this blue shape** (gesture)**?** (Answer: I) —**Orientations** Display shape C from the blue Shape Set turned at 45 degrees ◇. Under it display shapes W, C, A, and B from the yellow Shape Set. Ask: **Which yellow shape** (gesture) **matches this blue shape** (gesture)**?** (Answer: C)		

Building Blocks • *Assessment*

Learning Trajectory Record: Geometry (Shapes)

Age	Descriptors	Questions	Date	Response
3	**Shape Recognizer—Typical.** Recognizes and names typical circle, square, and, less often, a typical triangle.	*Question 2.* Use Shape Sets for the following questions: Show the following shapes to the child and ask: **What is this called?** Circle: C Square: S Triangle: D		
	Shape Matcher—More Shapes. Matches a wider variety of shapes with same size and orientation. **—Sizes and Orientations.** Matches a wider variety of shapes with different sizes and orientations.	*Question 3.* Use Shape Sets for these questions. Display shape M from the blue Shape Set, with the long side horizontal at the top Under it, display shapes B, I (△) W, and Z1 from the yellow Shape Set. Ask: **Which yellow shape** (gesture) **is the same shape as this blue shape** (gesture)**?** (Answer: I)		

Child's Name _____

Learning Trajectory Record: Geometry (Shapes)

Age	Descriptors	Questions	Date	Response
4	**Shape Recognizer—Circles, Squares, and Triangles +.** Recognizes some less typical squares and triangles, and may recognize some rectangles, but usually not rhombuses (diamonds).	*Question 4.* Use Shape Sets for these questions. Show the following shapes to the child and ask, **"What is this called?"** Square: C, turned ◇ Triangles: W, X, Y		
	Constructor of Shapes from Parts—Looks Like *Representing.* Uses manipulatives representing parts of shapes, such as sides, to make a shape that "looks like" a goal shape.	*Question 5.* Use straws or stirrers. Ask, **We're going to use these straws to make shapes. Can you make a triangle using some of the straws?** Check whether the children can at least approximate a triangle, such as the following: ◁		

Building Blocks • Assessment

93

Learning Trajectory Record: Geometry (Shapes)

Age	Descriptors	Questions	Date	Response
5	**Shape Recognizer—All Rectangles.** Recognizes more rectangle sizes, shapes, and orientations of rectangles. Correctly names these shapes as rectangles.	*Question 6.* Use Shape Sets for these questions. Show the following shapes to the child and ask: **What is this called?** Rectangles: T, U, V, Z1		
	Side Recognizer *Parts.* Identifies sides as distinct geometric objects.	*Question 7.* Use Shape Set triangle G for these questions. Show the shape to the child and ask: **Show me the sides of this shape.**		
	Angle (Corner) Recognizer *Parts* Recognizes angles as separate geometric objects.	*Question 8.* Show shape G again and ask: **Show me the angles of this shape.**		
	Shape Recognizer—More Shapes. Recognizes most familiar shapes and typical examples of other shapes, such as hexagon, rhombus (diamond), and trapezoid.	*Question 9.* Display all the Shape Sets shapes of one color in random order. Name the following shape categories one at a time, and ask: **Can you find all the _____?** Shape categories: Rectangles, Triangles, Trapezoids, Rhombuses, Hexagons		

Building Blocks • *Assessment*

Child's Name

Learning Trajectory Record: Geometry (Shapes)

Age	Descriptors	Questions	Date	Response
6	**Shape Identifier.** Names most common shapes, including rhombuses, without making mistakes, such as calling ovals circles.	*Question 10.* Use Shape Sets for these questions. Show the following shapes to the child, and ask: **What is this called?** Shapes: Rectangles, Triangles, Trapezoids, Rhombuses, Hexagons		

Learning Trajectory Record: Composing Shapes

Materials: Pattern Block Puzzles 7, 21, 26, 32 and Pattern Blocks

For each puzzle, the directions are the same.

1. Give the child the puzzle and the Pattern Blocks, randomly mixed in front of them.
2. Say: **Use Pattern Blocks to fill this puzzle. Put them together with full sides touching.**

Age	Descriptors	Questions	Date	Response
3	**Pre-Composer.** Manipulates shapes as individuals, but is unable to combine them to compose a larger shape.	*Question 1.* For Pre-Composer through Piece Assembler use puzzle number 7 and the pattern blocks.		
4	**Piece Assembler.** Fills simple outline puzzles using trial and error.	At the Piece Assembler level, the child is able to match most of the shapes, but may have 1–2 gaps and "hangovers."		

Building Blocks • Assessment

Child's Name _____

Learning Trajectory Record: Composing Shapes

Age	Descriptors	Questions	Date	Response
5	**Picture Maker.** Fills "easy" outline puzzles that suggest the placement of each shape (but note below that the child is trying to put a square in the puzzle where its right angles will not fit). Uses trial and error and does not anticipate creation of new geometric shape. Chooses shapes using "general shape" or side length.	*Question 2.* Use puzzle number 21. To succeed at this level, children have to completely and accurately cover the puzzle (with no gaps or "hangovers"). Children may use trial and error to do so.		
	Shape Composer. Composes shapes with anticipation ("I know what will fit!"). Chooses shapes using angles as well as side lengths. Rotation and flipping are used intentionally to select and place shapes. In the outline puzzle below, all angles are correct, and patterning is evident.	*Question 3.* Use puzzle number 26. Children solve the puzzle systematically, immediately, and confidently. They appear to search for "just the right shape" that they "know will fit" and then find and place it.		

Learning Trajectory Record: Composing Shapes

Age	Descriptors	Questions	Date	Response
6	**Substitution Composer.** Makes new shapes out of smaller shapes and uses trial and error to substitute groups of shapes for other shapes to create new shapes in different ways. Substitutes shapes to fill outline puzzles in different ways.	*Question 4.* Use puzzle number 32. As with the previous, children solve the puzzle systematically, immediately, and confidently. They appear to search for "just the right shape" that they "know will fit" and then find and place it. For this level, check whether children fill the hexagons in different ways. If not, have them complete the puzzle, then challenge them to do it a different way. (You may have to put the yellow hexagons away.)		

Child's Name _____

Learning Trajectory Record: Comparing Shapes

Materials: Shape Sets

Age	Descriptors	Questions	Date	Response
4	**Part Comparer.** Says two shapes are the same after matching only one side on each. "These are the same" (matching the two sides).	*Question 1.* For Part Comparer through to Congruence Determiner Use Shape Sets for these questions. Display both colors—blue on one side and yellow on the other. Ask: **Find some yellow shapes that match the blue shapes exactly.** Once the child has attempted 3–4 matches, tell them they are done.		
	Some Attributes Comparer. Looks for differences in attributes, but may examine only part of shape. "These are the same" (indicating the top halves of the shapes are similar by laying them on top of each other).			
5	**Most Attributes Comparer.** Looks for differences in attributes, examining full shapes, but may ignore some spatial relationships. "These are the same." ☐ ☐			
6	**Congruence Determiner.** Determines congruence by comparing all attributes and all spatial relationships. Says that two shapes are the same shape and the same size after comparing every one of their sides and angles.			

Learning Trajectory Record: Motions and Spatial Sense

Materials: Shape Sets and geometric puzzles (other shape manipulatives optional)

Age	Descriptors	Questions	Date	Response
4	**Simple Turner.** Mentally turns object in easy tasks. Given a shape with the top marked with color, correctly identifies which of three shapes it would look like if it were turned in different ways before physically moving the shape.	Give the child Shape Sets and/or geometric puzzles to explore, and observe how he or she manipulates them. *Question 1.* Show two of the same shape, and ask: **Are these the same? Show me.**		
5	**Beginning Slider, Flipper, Turner.** Uses the correct motions, but not always accurate in direction and amount. Knows a shape has to be flipped to match another, but flips it in the wrong direction.			

Building Blocks • Assessment

Child's Name _____

Learning Trajectory Record: Motions and Spatial Sense

Age	Descriptors	Questions	Date	Response
6	**Slider, Flipper, Turner.** Performs slides and flips, often only horizontally and vertically, using manipulatives. Performs turns of 45, 90, and 180 degrees. Knows a shape must be turned 90° to the right to fit into a puzzle.	As the child completes geometric puzzles, hopefully increasing in difficulty, work on a puzzle of your own and pretend to struggle with a few shapes. *Question 2.* **Can you help me complete this puzzle? How?** Tell the child to finish the puzzle, and observe his or her work. *Question 3.* **What do you think of these shapes? Tell me how you got them to fit. How did you know how to move them?**		
7	**Diagonal Mover.** Performs diagonal slides and flips. Knows a shape must be turned or flipped over an oblique line (45° orientation) to fit into a puzzle.			
8	**Mental Mover.** Predicts results of moving shapes using mental images: "If you turned this shape 120° it would be just like this one."			

Learning Trajectory Record: Patterning

Materials: Pattern Blocks

Age	Descriptors	Questions	Date	Response
3	**Pattern Recognizer.** Recognizes a simple pattern. "I'm wearing a pattern" about a shirt with black, white, black, white stripes.	Informal observation is a good strategy to see whether children can recognize patterns. You can also show a pattern and ask what they see.		

Child's Name _____

Learning Trajectory Record: Patterning

Age	Descriptors	Questions	Date	Response
4	**Pattern Fixer.** Fills in missing element of pattern, first with ABAB patterns.	*Question 1.* Lay out the ABABAB pattern of pattern blocks (red square, green triangle...six blocks in all) with one missing (ABAB_BAB). In front of the child, lay out six red squares and six green triangles in random arrangement. Say: **I made a pattern with these blocks but one block is missing. Can you fix this pattern?**		
	Pattern Duplicator AB. Duplicates ABABAB pattern. May have to work close to the model pattern. Given objects in a row, ABABAB, makes their own ABBABBABB row in a different location.	*Question 2.* Lay out the ABABAB pattern of pattern blocks (red square, green triangle...six blocks in all). In front of the child, place six red squares and six green triangles in random arrangement. Say: **I made a pattern with these blocks. Please make the same kind of pattern here using these cubes** (gesture in front of the child).		
	Pattern Extender AB. Extends AB repeating patterns. Given objects in a row, ABABAB, adds ABAB to the end of the row.	*Question 3.* Lay out the ABABAB pattern of pattern blocks (red square, green triangle...six blocks in all). In front of the child, place six red squares and six green triangles in random arrangement. Say: **I made a pattern with these blocks. Can you finish my pattern here the way I would** (gesture to the right of the pattern)?		

Building Blocks • *Assessment*

Learning Trajectory Record: Patterning

Age	Descriptors	Questions	Date	Response
4	**Pattern Duplicator.** Duplicates simple patterns (not just alongside the model pattern). Given objects in a row, ABBABBABB, makes their own ABBABBABB row in a different location.	*Question 4.* Lay out the ABBABBABB pattern of pattern blocks (red square, green triangle, green triangle…nine blocks in all). In front of the child, place six red squares and six green triangles in random arrangement. Say: **I made a pattern with these blocks. Please make the same kind of pattern here, using these cubes** (gesture in front of the child).		
5	**Pattern Extender.** Extends simple repeating patterns. Given objects in a row, ABBABBABB, adds ABBABB to the end of the row.	*Question 5.* Lay out the ABBABBABB pattern of pattern blocks (red square, green triangle, green triangle…nine blocks in all). In front of the child, place six red squares and six green triangles in random arrangement. Say: **I made a pattern with these blocks. Can you finish my pattern here the way I would** (gesture to the right of the pattern)**?**		
6	**Pattern Unit Recognizer.** Identifies the smallest unit of a pattern. Given objects in an ABBABBABB pattern, identifies the core unit of the pattern as ABB.	*Question 6.* Show a tower in an ABCABC pattern (red, white, blue…six cubes in all). Place *unconnected* cubes, three red, three white, and three blue, in front of the child. Say: **I want to make my tower bigger and keep the same pattern but I don't have any more small towers. Can you make a small tower with these pieces and keep the same pattern?**		

Building Blocks • *Assessment*

Child's Name _____

Trajectory Progress Chart: Number

Age Range	Counting	Comparing and Ordering Number	Recognizing Number and Subitizing (instantly recognizing)	Composing Number (knowing combinations of numbers)	Adding and Subtracting	Multiplying and Dividing (sharing)
1 year	Pre-Counter Chanter				Pre +/−	
2	Reciter	Object Corresponder Perceptual Comparer First-Second Ordinal Counter	Small Collection Namer			Nonquantitive Sharer
3	Reciter (10) Corresponder	Nonverbal Comparer of Similar Items (1–4 items) Nonverbal Comparer of Dissimilar Items	Nonverbal Subitizer Maker of Small Collections		Nonverbal +/−	Beginning Grouper and Distributive Sharer
4	Counter (Small Numbers) Producer (Small Numbers) Counter (10)	Matching Comparer Knows-to-Count Comparer Counting Comparer (Same Size)	Perceptual Subitizer to 4	Pre-Part-Whole Recognizer	Small Number +/−	Grouper and Distributive Sharer
5	Counter and Producer (10+) Counter Backward from 10	Counting Comparer (5) Ordinal Counter	Perceptual Subitizer to 5 Conceptual Subitizer to 5+ Conceptual Subitizer to 10	Inexact Part Whole Recognizer Composer to 4, then 5	Find Result +/− Find Change +/− Make It N +/−	Concrete Modeler ×/÷

Trajectory Progress Chart: Number

Age Range	Counting	Comparing and Ordering Number	Recognizing Number and Subitizing (instantly recognizing)	Composing Number (knowing combinations of numbers)	Adding and Subtracting	Multiplying and Dividing (sharing)
6	Counter from N (N+1, N−1) Skip Counter by 10s to 100 Counter to 100 Counter On Using Patterns Skip Counter Counter of Imagined Items Counter On Keeping Track Counter of Quantitative Units Counter to 200	Counting Comparer (10) Mental Number Line to 10 Serial Orderer to 6+	Conceptual Subitizer to 20	Composer to 7 Composer to 10	Counting Strategies +/− Part-Whole +/− Numbers-in-Numbers +/−	Parts and Wholes ×/÷
7	Number Conserver Counter Forward and Back	Place Value Comparer Mental Number Line to 100	Conceptual Subitizer with Place Value and Skip Counting	Composer with Tens and Ones +/− Fact Fluency to 20	Deriver +/−	Skip Counter ×/÷
8+		Mental Number Line to 1000s	Conceptual Subitizer with Place Value and Multiplication		Problem Solver +/− Multidigit +−	Deriver ×/÷ Array Quantifier Partitive Divisor Multidigit ×/÷

Building Blocks • Assessment

Trajectory Progress Chart: Geometry

Age Range	Shapes	Composing Shapes	Comparing Shapes	Motions and Spatial Sense	Length Measuring	Patterning	Classifying and Analyzing Data
2	Shape Matcher—Identical —Sizes —Orientations					Pre-Explicit Patterner	Similarity Recognizer Informal Sorter
3	Shape Recognizer—Typical Shape Matcher—More Shapes —Sizes and Orientations —Combinations	Pre-Composer Pre-Decomposer	"Same Thing" Comparer		Length Quantity Recognizer	Pattern Recognizer	Attribute Identifier
4	Shape Recognizer—Circles, Squares, and Triangles+ Constructor of Shapes from Parts—Looks Like	Piece Assembler	"Similar" Comparer Part Comparer Some Attributes Comparer	Simple Turner	Length Direct Comparer	Pattern Fixer Pattern Duplicator AB Pattern Extender AB Pattern Duplicator	Attribute Sorter
5	Shape Recognizer—All Rectangles Side Recognizer Angle (Corner) Recognizer Shape Recognizer—More Shapes	Picture Maker Simple Decomposer Shape Composer	Most Attributes Comparer	Beginning Slider, Flipper, Turner	Indirect Length Comparer	Pattern Extender	Consistent Sorter
6	Shape Identifier Angle Matcher	Substitution Composer Shape Decomposer (with Help)		Slider, Flipper, Turner	Serial Orderer to 6+ End-to-End Length Measurer	Pattern Unit Recognizer	Exhaustive Sorter Multiple Attribute Sorter
7	Parts of Shapes Identifier Constructor of Shapes from Parts—Exact	Shape Composite Repeater Shape Decomposer with Imagery	Congruence Determiner Congruence Superposer	Diagonal Mover	Length Unit Relater and Repeater		Classifier and Counter List Grapher
8	Shape Class Identifier Shape Property Identifier Angle Size Comparer Angle Measurer Property Class Identifier Angle Synthesizer	Shape Composer— Units of Units Shape Decomposer with Units of Units	Congruence Representer	Mental Mover	Length Measurer Conceptual Ruler Measurer		Multiple Attribute Classifier Classifying Grapher Classifier Hierarchical Classifier Data Representer